The
Cloud Forest

by
Gail Bornfield

Forward

The Cloud Forest is set in Costa Rica where clouds dance on treetops and rest on volcanoes. Plants and insects snuggle together in trees which become their homes. Marcos wakes to the bellowing of the howler monkeys and songs of passing birds. The forests have been cut. Birds and animals are becoming extinct. What can be done to save them? Can a group of kids help? Will the volcano erupt?

Dedication

The Cloud Forest is dedicated to my grandson, Caleb Young. He is a young person who loves the world of nature, the beauty of animals and the lives they lead.

Books by Gail Bornfield

Big Blue

Riding the Wind

River of Fire

The Lonely Sea Lion

The Orphan Elephant Herd

The Winds of Change

What Could It Be?

Chapter 1

Marcos ran to the window. It was true the volcano was smoking. He had never known it to erupt, but it was smoking. What would happen now? It was time to get ready for school. What if it erupted on his way to school? He had seen pictures of volcanoes shooting molten lava into the sky. Would he have to move? Was there time to pack?

He ran downstairs. "Papa, the volcano is smoking. What will happen now?"

"Calm down, it is only smoking. Volcanoes do that sometimes. Nothing is going to happen today. We will listen to the news reports to hear what the scientists are saying about it. Now, let's eat breakfast and get off to school."

Marcos couldn't take his eyes off the plume of smoke. It was just a single stream of smoke going straight up into the air. Maybe the teacher would talk about it in science class. He walked faster.

When he arrived at school, everyone was talking about it. There was a mixture of excitement and fear. Some of the teachers remembered the eruption that happened in the year 2010. They were concerned.

Marcos lived in Arenal at the base of the volcano. It was near the Monteverde Cloud Forest in Costa Rica. He had taken the volcano for granted, never thought much about it. He could see the lava paths streaming down the one side, but they were old and had formed into rock. People had talked about the eruption, but everyone thought that it would not happen again for many years. It had been 400 years between the last eruptions.

The science class had been studying native birds, but the teacher decided to spend today talking about volcanoes. Mrs. Carrillo began the class, "How many of you saw the plume of smoke from the volcano this morning?"

Everyone raised their hands. "It is kind of scary. I remember when the volcano erupted a few years ago. It was both beautiful and frightening. We were very lucky, because no one was hurt. There were some homes lost on the other side of the volcano where the lava runs are. The volcano was good to us, because it gave us lots of warning before it erupted. So, we should be aware of it, but we shouldn't be too afraid."

"How do you know when it is going to erupt?" someone asked.

"We count on the Volcanic Activity Alert System to tell us when that will happen. They have a system that reports the activity level of the volcano. So, most of the time it is 'normal' which is what it was yesterday. Nothing was happening. It looked like a mountain. Today, it might be at an 'advisory level' which means there is unrest within the volcano. There are two more levels of alert. The third level is 'watch' which means that there is unrest with an increased potential for eruption. And, the final level is 'warning' which means that eruption is imminent. It could happen any time. The news reporters will keep us updated on the level of alerts. We need to pay attention to those reports so that we know what is happening with the volcano. What do you know about volcanoes?"

"It looks like fire when it explodes. It's dangerous."

"It runs down the side and looks like a river of fire."

"Yes, it does look like fire when it explodes, and the lava flows down the side of volcano. There are types of volcanoes. Our volcano is called a 'composite volcano' which means that it is made of many layers of volcanic rocks. Most of the time, these rocks include cinders, ash, and lava. The cinders and ash pile on top of each other, and the lava flows on top of them where it cools and hardens. This process creates layers of rock."

"What causes it to explode?"

"That is a good question, Marcos. In order to understand how volcanoes erupt, we need to know a few things about the earth. The earth is made up of three layers. The first layer is the 'crust' which is the outer layer. It is made up of large slabs of rock called plates. It is about 18 miles thick, and we build our homes on the 'crust'. The second layer is called the 'mantle', and it is about 1800 miles thick. Between the 'crust' and the 'mantle' is a substance called magma which is composed of rock and gases. Finally, the center of the earth is called the 'core'.

"The plates located in the crust fit together like a large puzzle. Sometimes, these plates move. When two of them collide, one plate slides on top pushing the other one down underneath it. The magma gets squeezed up between these two plates from the pressure and is forced out of its location. This pressure on the magma can result in an eruption.

"We are lucky, because the smoke is actually a stream of gases escaping which is relieving some of the pressure inside the volcano. I hope that all makes sense."

"So, we are safe?" Marcos asked. "The smoke is a good thing and means that the volcano won't erupt. Is that right?"

"Not totally. The smoke is both a safety hatch and a warning. So, we need to listen to what the volcanologists are saying. Those are the people who study volcanoes. They will have reports on the news. You can get them on any of your WiFi devices. Our time is nearly over. We will get back to our lessons on native birds tomorrow. Try not to worry about the volcano tonight."

Marcos wasn't sure how safe it was, but he felt a lot better. Mrs. Carrillo knew a lot about science. He walked home with his friends. They kept their eye on the volcano and vowed to get the news as soon as they got home. The news report said that the volcanic activity level was at 'advisory'. Marcos was glad that Mrs. Carrillo had explained those levels to the class. He knew that scientists were watching the volcano, but he still felt uneasy and a little scared.

The next morning, the volcano was still smoking. Marcos had breakfast and walked to school where Mrs. Carrillo was ready with the lesson for the day on native birds. He couldn't imagine life without birds. They woke him up in the morning and sang him to sleep at night. They were everywhere. He loved the

Scarlet Macaws which had nearly become extinct. The biologists were working to save them and reintroduce them to different parts of country.

When Marcos arrived home from school, his father, Pedro, was waiting for him. "Marcos, we are going to a lecture later. We will pick up Grandpa. He will go with us too. The lecture is important for us to hear. A biologist, Molly, from the United States is going to talk about the Cloud Forest and the loss of wildlife. The Cloud Forest is a special place. I hope that a lot of people will attend. This is important for everyone."

"Papa, did you hear anything about the volcano?"

"It is at an advisory level. There is nothing to worry about."

Marcos still felt a little worried. It wasn't every day that the volcano sent up streams of smoke. He felt uncertain about the whole thing. It didn't feel safe.

Marcos knew that some of the birds had disappeared and that some were in danger of extinction. These were listed on the endangered species list. Pedro was a tour guide for one of the big tour companies in the country. He liked to share information about the country with his tour groups. He traveled all over the country with his work. He had seen firsthand how wildlife patterns had changed over the years.

Julio, Marcos' grandfather, was a farmer. He remembered when Costa Rica was covered by rain forests, and the golden toad was everywhere. When the sloths were considered a nuisance. When the Scarlet Macaws could be seen flying across the forests. He knew things had changed. The wildlife that he grew up with was disappearing.

Pedro stopped the truck to pick up Julio. "Are you ready?"

Julio hurried to the truck. It was quite a long ride to the hotel where the lecture would be held. Although Marcos had been learning to identify some of the birds and animals, he did not know much about them. He did not know much about the Cloud Forest either. He wondered what the lady would talk about.

"Grandpa, did you hear anything about the volcano?"

"Not really, Marcos. I do know that it is still smoking. I don't think we will have to worry for awhile yet, but we need to keep an eye on it."

They finally arrived at the hotel and went in to find a seat. There were about 20 people already there. Pedro was glad to see so many showing up for the lecture. He knew that the more people who came, the better it would be for the wildlife.

Molly began the presentation by welcoming everyone. She shared that she came from the United States to study the Cloud Forest. She loved Costa Rica so much that she stayed. She had been here for ten years.

"I am glad to see so many people. I am hoping that the volcano will choose not to share his voice with us tonight." Molly said, then laughed with the others.

"The Cloud Forest is a special place," she continued. "It is home to many different species of birds, mammals, insects, amphibians, and reptiles. There are some species that are endemic to the forest. That means that it is the only place on earth they live. There are other species that have become extinct, because they lost their habitat. I always think of the Golden Toad. It took such a short time to go from many to none, just a few short years.

"We have learned a lot about caring for our plants and animals in the past few years. We saved the Scarlet Macaw from extinction. We have reintroduced it back into the forest and to the Pacific Coast.

"We are in danger now of losing the three-wattled bellbird to extinction. They are migratory, meaning that during certain times of the year they move from one place to another. They begin in southern Nicaragua or northeastern Costa Rica and come to our Cloud Forest where they mate and hatch their young.

They are usually with us from March through September each year. Then, they return to where they originated.

"Three-wattled bellbirds have one of the loudest bird calls in the world. People can hear it up to a kilometer or two-thirds of a mile away. The male has striking black and white plumage with three black wattles dangling from his beak. The females' coloration allows them to blend into the forest. The females are secretive in their behavior.

"The problem is that the habitats that it needs to get here to the Cloud Forest no longer exist. It has made the journey difficult and sometimes deadly. The birds have a hard time finding food. Bellbirds like to eat wild avocados. But there are very few avocado trees left in their migratory path. With the loss of the avocado trees to deforestation, their numbers have been declining since 1997.

"We think we know how to remedy this situation, how to help this bird and many others. We must develop a biological corridor for them. A biological corridor consists of pieces of land that have been given to reforest areas that were once forested. It will give the birds and other wildlife a highway to get here to the Cloud Forest and to return to their other habitats. The birds are unable to cross large expanses of open space. They must have trees on which to rest and feed.

"Some of you may not think that these little birds are important. But, they are a species of the earth and

perform a function of pollination necessary to maintaining some of our plant life. If the bird becomes extinct, it is very possible that we will lose a plant species with them. If it happens over and over, we will lose the forest.

"The biological corridor is important to the future of the country. We have a plan. If the farmers will allow us to use a small ribbon of space across their land, we can plant trees traditionally found on the land. Then the birds and other wildlife will have a better chance of surviving upon their return. Are there any questions at this point?"

"If we plant the trees, how do you know that the wildlife will return?" a farmer asked.

"That is a good question. We have conducted experiments on a small scale. We have planted a corridor between two forests to see if the wildlife would use it. And, they did. So, we are confident that if we plant the trees, the wildlife will return to the forests," Molly explained.

"Where are the trees coming from? Are you growing them?" someone asked.

"That is a problem. We do not grow them. We must buy them, and they are expensive. We do not have enough money for all the trees that we need. We are hoping you can help us pay for some of these trees," Molly said.

"We do not have any money to buy trees," a farmer said. "We are small farmers. We barely make a living for our families. How can we afford this?"

"I understand that this is an expense not affordable for everyone. But we cannot do it alone. We need your help to build the biological corridors necessary to support wildlife," Molly said.

"How can we help?" Pedro asked.

"We must find a way to raise money. Perhaps fundraising activities would help. We will be asking for donations. We are at the very beginning of this

- -

project. At this point, we need to know whether you consider biological corridors to be important to the future of the country," Molly said.

"I think it is very important," Julio said. "I remember animals and butterflies, birds and toads that are gone. They will never come back. They are gone from the earth. They are extinct. My grandson will never see them or hear their songs. This is very important to us."

"How will we know where to plant the trees if we get them?" asked another farmer.

"We will help you find a good location," Molly said. "We will be asking the farmers to help us with this. We want to use land that is not as good for you to plant crops or graze cattle. Thank you all for coming and sharing. Please leave your addresses, emails, and phone numbers. I will be in touch with you as we move forward with this project."

As they left the meeting and got into the truck, Pedro asked Marcos, "What did you think of the talk?"

"I don't know for sure, but I don't want any more birds or animals to go extinct," Marcos replied.

"Me neither, I've watched too many species be lost to the earth. It is very sad when we don't see them. You expect particular birds to come at a certain time of the year, but they don't come. I miss their songs and beautiful colors," Julio said.

"Let's get home. We can sleep on it. We'll see what tomorrow brings," Pedro said.

The next morning, Marcos was off to school. He couldn't stop thinking about the need for the biological corridor. But, what could he do? How could he help? He was just a kid.

During science class, Mrs. Carrillo shared information about the three-wattled bellbird. She told the class about how unusual it was. She told them it was on the endangered species list. "It's most unique feature is its call which we can hear up to one kilometer. Its call sounds like a squeak-bonk. It is one of the loudest birds in the world. Bellbirds live in the Monteverde Cloud Forest between March and September. I want to tell you a little bit about the Cloud Forest.

"As you know, the Cloud Forests are located on mountains in tropical areas like Costa Rica. They are cloud covered most all of the time. The forest seems to rest within a bank of clouds. It is comprised of deep green foliage formed from epiphytes such as mosses, ferns, and bromeliads. Epiphytes are plants that live on other plants including trees. These plants collect their moisture from fog. The clouds hover around the canopy of the forest condensing into droplets of water on the leaves of the tress which drip onto the plants below. So, it doesn't actually rain.

"The Cloud Forest is such a unique environment, and there are only a few within the whole world. We want to keep ours healthy and filled with wildlife. Do you have any questions or thoughts about the wildlife or the forest?"

Marcos shared what he had learned at Molly's talk the night before. He told them about the biological corridor and the need for money to buy the trees. "I want to help. But, I don't know how I can."

"It sounds like money is the biggest need. Who will plant the trees if they do buy them?" Mrs. Carrillo asked.

"She didn't say. I don't know," Marcos said.

"So, there are two problems. The first is the need for money to buy the trees, and the second one is the labor to plant them. It might be that if enough of you are interested we could form an after school club to work on a plan to help. How many of you would be interested in that?" Mrs. Carrillo asked.

About half of the class raised their hands.

"I am so glad we have so much interest. Marcos – what do you think about the idea of a club? What is the first thing we should do?" Mrs. Carrillo asked.

"I think we should ask Molly if she could come to talk to us about the biological corridor and the trees. We need a name for the club too," Marcos said.

"What a great idea Marcos. Let's invite Molly to come to our first meeting. I can ask her. I need all of you to think of a name for our club. Would Wednesday after school work for a meeting for all of you? Please raise your hands if it works."

The hands went up. The meeting was set for Wednesday after school.

"I will get a note ready for your parents to give them some information about the club and the meeting times. Now, it is time to move on to history."

The students put away their science materials and got out their history books. There was an excitement in the air. The students were smiling and whispering about the biological corridor and the trees. Mrs. Carrillo smiled at their enthusiasm and gave them a few minutes to share quietly with each other.

Marcos was excited about the club and the idea of doing something to save the forests. When he got home, he told his father about the club. "I'm sure that Mrs. Carrillo had been thinking about starting this club for a long time. It didn't have anything to do with what you shared from Molly's talk. It was just a coincidence. I am glad that Mrs. Carrillo is doing something to help you kids learn more about preserving the forests."

"Me too," Marcos went to his room and started his homework. He almost wished that he hadn't told his father. His enthusiasm was gone. He put his head down in his arms on the desk. After awhile, he lifted his head and began paging through his books. A tear rolled down his cheek. He wished his father could be proud of him.

Chapter 2

Mrs. Carrillo was pleased to see so many students at the first meeting. She called the meeting to order. Then she asked Marcos to share what he knew about the need for biological corridors.

"My grandpa says there are a lot of birds, butterflies, toads and plants that used to be here that are gone now. They are gone forever. We will never get a chance to see or hear them. I don't want that to happen to the three-wattled bellbird or any other bird. Molly said that it would help the birds if trees were planted on the farms. I want to help."

"Thank you, Marcos. Now, I would like to introduce Molly. She works with a special project to help preserve the forests. She is here today to tell us about the problem and her project."

"Thank you for inviting me to your meeting," Molly began. "The best way to explain this is to think about how we get to school every day. Each of you has a path you take to get here. Some of you come a long way, so you use the bus on the highway. Others of you are able to walk from your homes.

"Let's think of the birds' paths to get home to the Cloud Forest. It is the same. They need a highway. But, their highway is built of trees and plants. Animals often use this highway too. If that highway is destroyed or doesn't exist anymore, the birds cannot get to the Cloud Forest. If they can't get to the Cloud Forest, they will die. That is why it is so important to plant the trees. They create the highway for the birds to travel, allowing them to migrate. The trees create a small forest which will provide food and resting places for them. I have talked a lot. Do any of you have any questions?"

"So, the birds need a road to the Cloud Forest. And, the road is made of trees. Is that right?" Adelina asked.

"Yes. That is exactly right," Molly said, smiling. "Are there other questions?"

"How can we help?" Carlos asked.

"That is a very good question. How can you help? Let's begin by sharing some ideas with each other. Does anyone have an idea about how we could help?"

"Maybe we could sell something."

"That is an idea. What do we have to sell?"

A few ideas were shared. Some of them sounded helpful.

"Remember, we must think about what can be done here in our area. We can't solve it for the whole country. What can we do here in our community to establish biological corridors?" Mrs. Carrillo prompted.

"We could talk with other people about the need for the highway and ask them to help us think about it," Marcos shared.

"Good idea, Marcos. We will share the ideas that we gain from talking with others at our next meeting. Now, let's all give Molly a big thank you for coming to visit us." The club members clapped loudly.

Marcos felt proud that Mrs. Carrillo thought his idea was a good one. He wished his father could feel proud of him.

At home that evening, Adelina explained the problem to her family. She shared the ideas from the club for raising money. Then, she asked them if they had any ideas that could help.

"Adelina, we understand the problem. But, we don't know how to help. Maybe you could tell us a little more about what you need," her mother said.

"We need money to buy the trees to build the biological corridor. Then we will need help to plant them."

"Okay. I understand now. One idea that comes to me is that the Flower Festival will be held in a couple of months. The club could have a booth at the festival. The parents could bake cookies, and club members could sell them. It would not be a lot of money, but it would be a start. What do you think?" Her mother asked.

"I like that idea. I could help you make the cookies. It would be fun. Thanks, Mom." Adelina gave each of her parents a big hug, as she went off to bed.

Carlos felt a little strange asking his parents for help. He didn't think they would be very interested in saving the birds. He knew they loved him, but he also knew they weren't too interested in his activities. They never asked him about anything he did. It was okay. He shared his stories with his friends. He had a good time with his parents when they did family stuff.

He thought of how his father liked to fish on the river. They would go together. There were so many birds and iguanas along the river. He really liked the caiman. They looked like crocodiles. The last time they went fishing; they found a nest of babies. It was the first time he had ever seen baby caimans. They were cute like all babies.

He wanted to help save the birds. So, he would have to ask them for their ideas. It was his responsibility as a member of the club. He would have to report back at the next meeting. He wanted to have something to share.

What would they say? He wondered. They may just say that it is silly to think that we can help save the birds. He prepared himself for the worst. The time had come to talk with them. He explained the problem to his parents.

"Mom, I need to ask you and Dad something. I have joined a club at school. We want to save the birds by planting trees to create highways for them to use during migration. We need to raise money. We need ideas about how to do that. Do you have any ideas that could help us?"

His parents just looked at each other and then at him. He had never asked them about anything at school. They felt flattered. They wanted to help him. His father spoke, "Carlos, we are pleased that you would ask us to help with this. Perhaps your mother and I could take a day to think about it. We could share our ideas with you tomorrow night. Would that be okay?"

Carlos was so surprised, he just grinned and said, "Yes!"

Marcos knew his father would have good ideas. He was a tour guide and traveled around the entire country. He saw many things and knew how to help in situations just like this. He always had lots of ideas, and he was interested in the Cloud Forest. Marcos felt sure that he would come up with good solutions. He explained the club and the need for money to buy the trees.

But when Marcos asked, his dad said, "No Marcos, I don't. This club is yours. You kids should be coming up with your own ideas, not going to your parents and others for help. You need to learn to solve your own problems. Do you understand? You need to come up with the ideas. Your mother and I are not going to do your work for you."

"I understand. I will try to think of something," Marcos said, disappointed.

"You need to do more than try. You need to come up with an idea."

Marcos stared at the floor as he walked to his bedroom. He didn't know anything about raising money. He rarely had any money. Sometimes his grandfather would give him some small change left in his pocket. As he changed into his pajamas and crawled into bed, tears rolled down his face.

What could he do? Maybe he could ask his grandfather. But, he might tell his father who would be even more disappointed in him. No, he had to think of something.

The next morning, Marcos woke to the bellowing of the howler monkeys. They were small in stature but loud in voice. They sounded like huge roaring lions. They began every morning at five thirty. As soon as they began chattering, the birds began chirping. It was a symphony of sounds in the morning. It was nature's alarm clock.

The sound of the monkeys always made him smile. He loved the morning sounds. It was the beginning of a new day. The monkeys and birds had already welcomed it. Now, the day was his to begin. He got dressed and checked the volcano to see if it had changed, which he did every morning. There were no changes. He joined his mother for breakfast. She had made rice and beans with fresh pineapple and papaya for him.

"Mom – do you have any sour soup juice left?" Marcos asked.

His mother got the juice for him. Sour soup juice was made from the sap of a local tree. It wasn't as sweet as some fruit juices. It was tart with just a hint of sweetness. It was a little like lime juice. He loved it.

"It's time to go to school. Your friends are probably waiting for you. You better hurry, Marcos. Have a good day. I'll see you tonight." She gave him a big hug good-bye.

As Marcos walked to school, sometimes friends joined him. Other times he walked alone. It was about a mile. He enjoyed the walk. There were many trees and plants along the way. He liked the machete tree. In spring the tree was cloaked in a bright orange veil of blossoms. The blossoms resembled the shape of a machete, which is how it got its name. He turned to the sound of the green macaw. Their squawk was easy to identify, but they were hard to see unless they were flying. Even then, they were visible only at a distance. He enjoyed knowing they were there. They were special to Costa Rica.

He needed to begin thinking about an idea to raise money. As he was walking, a neighbor joined him. Martin was older and in high school. Marcos told him his story along with the problem.

After some thought and a bit more conversation, Martin said, "I am a member of the Biology Club at the high school. I could bring it to their attention. We could make it our mission for the year. We could use the slogan 'Bring back the trees!' What do you think?"

"That would be great! Do you really think the other kids in high school would want to help?"

"I think so. Many of us want to preserve the forests and take care of the wildlife. Our teachers talk to us about it in science class. Yesterday, she was telling us about the blue-jean frog on the other side of the country. It is so cool. It is about the size of a quarter with blue legs and red eyes. If I can get students to come, will you come and talk about your club? Maybe the groups could work together and help each other."

"What would they do?"

"I think the club members would be willing to plant the trees. We have a lot of members. They could share the work."

"That really sounds good. I could talk to the club members. I feel better. Thank you for helping me. I can hardly wait until Wednesday comes. I can tell Mrs. Carrillo and the other kids that our idea has grown. We have help to plant the trees. They will be so excited," Marcos said.

"I hope we can plant a few trumpet plants. The ants build condos in them. They look like rings around the tree, but each ring is a separate and unique ant colony. Nature is so cool!"

Chapter 3

The next day, Carlos was excited to hear what ideas his parents had come up with to help raise money to build the biological corridors. He lived about a mile from the school, so he walked and ran and walked again trying to hurry home.

He arrived home before his parents. So grateful for their interest, he decided to fix them a snack. He had watched his mother cut up fruit hundreds of times. She even let him help once in awhile. He was sure he could do it. His parents loved their coffee. The coffee they liked best was raised on a nearby coffee plantation, which also raised bananas. It was Arabica coffee.

Proud of their coffee, Costa Ricans only raise the very best. It is the law. They export coffee to the United States and other countries.

The owners of the coffee plantation still raise and harvest the coffee beans in the traditional ways. They start their plants as seeds. It takes three years for the plant to produce coffee beans. When the beans are ready, the farm workers are brought in to pick them.

The beans are picked by hand. They are placed in a basket that is tied around the waist of the worker. When the basket is full, they are taken to the warehouse where they are placed into a giant sorter which separates the best beans from the others using a water process. The best beans sink, the other beans float.

Once the separation process is complete, the beans are taken outside and spread on large cement slabs where workers rake and push them into a single layer for drying. They lay in the sunshine for several days. While they are there, the workers turn the coffee beans over so that they will dry evenly on all sides.

At the end of the drying process, they are pushed into large piles in preparation for roasting or shipping. Those that are being shipped to other countries will be bagged and shipped. Those that will be used in Costa Rica are roasted. Once they are roasted, the beans can be ground and put into the coffee maker. Carlos' parents liked a medium roast.

His mother had let him help her make coffee a couple of times. He remembered her telling him to fill the coffee up to the line, turn on the machine, and to make sure there was water in it.

He felt good about doing something special for his parents. He heard the car drive up. His parents were coming toward the house. He ran to the door, opening it and greeting them with a big smile.

"Mom, Dad welcome home. Please come into the kitchen. I have something special for you." He smiled walking toward the treats.

His parents looked intently at each other and followed him quietly. "Carlos – did you make these for us? I can't believe it. I didn't know that you knew how to do all of this. This is unbelievable. Thank you," his mother said giving him a big hug and kiss on the cheek.

"You work hard. I wanted to do something special. I hope you like it."

"We like it Carlos," his mother shared as they sat down to enjoy their snack.

"Have you thought of an idea to help us with the trees?" Carlos asked.

"We have given it a lot of thought," his father said. "Your mother and I have discussed it, and we have two ideas. First, you could do odd jobs to earn a little money that you could donate for the trees."

"What would we do? They won't let us work at the plantations. We are not old enough."

"That is true. Maybe you could do a little work for us or for your grandparents, and maybe the neighbors," his father said.

"What kind of work?" Carlos asked.

"I was thinking of chores like pulling weeds, cleaning up the trash, helping with the pets, or washing dishes. Does that sound reasonable?"

"I guess," Carlos said. He didn't sound convinced. He already had a few chores. When would he do this? He had homework too. And he needed time to be with his friends. Where could he find time for more work?

"I can see that idea is not to your liking Carlos," his father said. "We have one more idea. You could take used cans and cut slits in the top to make them ready for coins. These could be placed in local stores where people could donate their change to help plant the trees. Each of the cans could have a label saying something like 'Bring back the trees'. Maybe the newspaper would do a little story about your project. It would help people to understand why their donation is important."

"I like that idea," Carlos said. "We can do that. Maybe all the stores in town would help us. We could make enough to buy a few trees. Thanks Dad! Thanks Mom! That is a great idea. I can hardly wait to tell the others."

"We are pleased to help. We are even more pleased that you like our idea. Come join us for some of this nice fruit. Soon you will be old enough to have coffee with us. By the way, this coffee is perfect. You are growing up quickly."

His mother suddenly realized that he wasn't a little boy any more. He had ideas. He understood things about the environment. He wanted to help. She felt a tear fall from her eye. Her little boy was growing up.

Chapter 4

Mrs. Carrillo convened the meeting of the club.

"Welcome to our second meeting. I am so glad that the volcano is still just smoking. It hasn't gotten any worse. We are very lucky. Now on another note, I am hoping that everyone spoke with their parents or other people and brought lots of ideas back with them. How many of you were able to speak with someone who was helpful with ideas?" Everyone raised their hand.

"That is very good. I am so impressed with all of you. Adelina, would you like to share with us first?"

"The Flower Festival is coming. So, we wondered if the club could have a booth at the festival. Everyone could bring cookies or cakes to sell. We could raise enough money for a few trees."

"That is a great idea, Adelina. Who would be willing to bring a cake or cookies?" Again, all the hands went up. "Great! Adelina will you help me make a call to the chair person for the festival. We will need to secure a booth for our club. There may be a small cost to it. We may have to ask that the fee be waived. So, Adelina, I will need you to be ready to explain the purpose of our booth. You will need to tell her what we will do with the money. Can you do that?"

- -

"I think I can." Adelina wasn't quite sure that she could, but she knew it was important to raise the money. She thought that she could talk about the trees.

"Who is next?" Mrs. Carrillo asked.

"I asked my parents to help me think of ideas," Carlos volunteered. "We thought it would be a good idea to make cans with slits in the top for spare change and put them on the counters at the stores. Labels on the cans could read, 'Bring back the trees!' Or something like that. Maybe the newspaper would do an article about our project. I think we need a name for our club."

"That's a great idea. And you are right, Carlos, we do need a name for the club. Let's finish with the ideas first. Then we can work on a name. I like the slogan. I know a reporter for the paper. Would you come with me to talk to her about the idea? Maybe you could have a can ready to show her. What do you think?"

Carlos wasn't eager to talk with a reporter. He didn't have a can made yet. He wasn't even sure how he was going to gather the cans, nor how he was going to put the slit in the top and the slogan on the side. But, he knew he needed to help. So, he said 'yes'.

It was time for Marcos to share. He felt a little strange, because his idea was so different from the others. And his parents had not helped him. But, he decided to share the idea that he and Martin had discussed.

"I have a friend who is a member of the Biology Club at the high school," he began. "He thought the high school kids might be willing to plant the trees if we buy them. He is going to tell them about our project and ask them if they would help. He thinks their club would be happy to work with ours."

"So, we have ideas for both problems – buying the trees and planting them," Mrs. Carrillo said, smiling. "That's great. It will be especially nice to collaborate with the Biology Club. They may have other ideas to share with us. We may be able to help them with other projects, too, although this is a big project and will take most of our time for quite awhile.

"I want to thank you all for your great ideas. We need to move forward. Carlos, could you get started on the cans? Adelina, contact everyone to find out if they can donate baked goods to the booth. Marcos, could you invite the Biology Club representative to our next meeting to share information about their activities? Now, we still need a name for our club. Are there any ideas?"

"We could be the 'Biology Club' at the elementary grades," a member shared.

"How about the 'Environment Club'?" another asked.

"Maybe, 'Save the Trees Club'?"

"Would 'Protect Nature Club' work?"

- -

"I know. We could just call it 'The Nature Club'. That would cover everything – the trees, the birds, the other wildlife," Marcos said.

"Yeah, I like that," someone agreed.

"Okay. I think we have a name," Mrs. Carrillo said. "We will be 'The Nature Club". Thank you for your good ideas. We have work to do. So, I will see you back here next week. But first, Adelina has a question for you."

"Is everyone willing to bring a cake or cookies to sell at the booth? Are you willing to work at the booth during the festival?"

Everyone agreed they were willing to help and bake. Adelina realized there still would not be enough baked goods for two or three days of selling. She needed a lot more. Who could she ask? Even if the whole class volunteered, it still would not be enough. She needed helpers and bakers.

"Adelina, you might want to think about who has baked goods that they might be willing to donate," Mrs. Carrillo said.

"Grocery stores and bakeries have a lot of baked goods. Do you think they might be willing to help us?" Adelina asked.

"I think you should ask them. Their donation would be for a good cause, and you will ask with a smile,"

Mrs. Carrillo said, demonstrating a smile. "Okay, time to go home. I'll see you all tomorrow in class."

"I will ask. Would somebody like to go with me?" Adelina asked.

"I'll go," Marcos volunteered.

"I could go too," Carlos said.

"Great! All three of you can go to each of the stores," Mrs. Carrillo agreed. "You'll need to work up a short presentation, so they understand why you are asking for the baked goods. You might want to ask permission to place the cans on the counter at the same time. Could the three of you stay after school tomorrow? We can work on a presentation for the stores."

They, of course, volunteered to stay after school to work on the presentation. Adelina gave a big sigh of relief. She wasn't sure how she would approach the stores for a donation and was glad to have someone to go with her. The three walked home together.

"The volcano is still smoking," Adelina said as they walked.

"It is not any worse than before," Marcos said. "The news reports say it is still at an 'advisory' level. Don't worry, Adelina, it will be all right."

"I just wish it would stop smoking," Adelina said. Not wanting to scare Adelina, Carlos and Marcos silently wished it would stop too.

Marcos arrived home just as his father pulled into the driveway. After exchanging their greetings and hugs, his father wanted to know how the project was going. Marcos explained what was going on in The Nature Club. He shared the ideas that the members had brought into the meeting.

"Marcos, that is wonderful. Everyone is contributing and thinking about the problem. They are getting involved. It is exciting. I think that if the club continues on this path, they will make a difference for wildlife in our area," Pedro said.

Marcos felt good. His father could be a bit hard on him at times, but it always felt good when he was happy about Marcos' activities.

"I am going over to visit Grandpa Julio. Would you like to come?" Pedro asked.

"Sure. I haven't seen him since the lecture."

It was about an hour drive to Julio's farm. When they arrived, Marcos shared the news from The Nature Club with him. Julio was excited to hear about all of the ideas surrounding the project.

"Marcos, I want to help," Julio said. "I could volunteer to work in the booth during the festival. Maybe I could bring pictures of some of the animals and birds that are extinct now. It would remind people that the problem is real and needs attention. We are losing our native species because of poor decisions made many years ago. I might have pictures of the golden toad. It seemed like one summer there were thousands of them, and the next summer there were none."

"Grandpa, that is wonderful. I will tell Mrs. Carrillo that you want to help at the booth. She will be glad, because we do not have enough people to help. We don't have enough baked goods either."

"Then, it is settled. I will help. I want you to come outside with me. I want to show you a plant that is blooming. We call it the rattlesnake plant, because the blossoms look like rattles on a snake. Isn't it a beautiful yellow? It grows wild in the jungles. But, I transplanted it to my garden. It seems to like it here."

"I like its name, Grandpa – the rattlesnake plant."

"Oh look, the hummingbirds are coming into the feeders. They are a bit territorial. Some of them will not allow others at the feeders until they have finished. It seems almost like a hierarchy of species. Some are more powerful than others, so they get to eat first. Their wings flap so fast that they are almost invisible."

"Let's walk over there. I planted a butterfly plant hoping that more butterflies would visit. And, look at that. It is a Blue Morph butterfly. I haven't seen one of those in quite awhile. When they sit and close their wings, they are brown. That protects them by allowing them to fit into the colors of the environment around them, like the trees and branches.

"When they fly, they open their wings which are a beautiful luminous blue. They spread them for the world to see and the sun to shine upon. Marcos, over there, it is an Owl butterfly dressed in brown and gold colors. We call it that because when its wings are open they look like owl eyes. It also has the outline of a snake on its lower wings. It is Costa Rica's largest butterfly. I love nature's creatures. I think they know. That is why so many of them visit my garden," Julio said, smiling.

"Time to go home Marcos," Pedro called.

"Okay, Papa. Thanks, Grandpa. I sure love your garden." Marcos gave Julio a big hug.

On the ride home, Marcos shared that Grandpa had volunteered to help at the festival.

"That's great, Marcos. Maybe a half day for a couple of days would work out for Grandpa. He isn't as young as he once was."

"I understand. I'll tell Mrs. Carrillo." It was a quiet drive home.

The next afternoon, Mrs. Carrillo met with the kids. "What should we include in the presentation? Remember it shouldn't take more than a few minutes to explain what we are doing. People are busy. They won't listen to us for very long."

"First I think we have to explain to them that we want to buy trees to save the birds and animals," Carlos offered.

"And, we should tell them about the biological corridor," Adelina shared.

"How can we do that in a few sentences?" Mrs. Carrillo asked.

"We could start with 'Lots of animals and birds have gone extinct in Costa Rica, because the forests are gone.' Would that work?" Carlos asked.

"That is a good start. What's next?" Mrs. Carrillo asked.

"We want to build a biological corridor with new trees to help migrating birds and animals," Marcos said.

"That will do it. Then you could add, 'We are hoping you will let us put this can on your counter near the cash register. People can put their change in it. We will use the money to buy trees for the biological corridor.' What about the baked goods?" Mrs. Carrillo asked.

"We could say, 'We will also have a booth at the Flower Festival. We will be selling baked goods. Could we count on a donation from you?' Would that work?" Marcos asked.

"Very good. What if they say 'yes'? What do you need to tell them?" Mrs. Carrillo asked.

"We should thank them for their donation. We could ask them what they are donating and tell them that we will pick it up the day before the festival," Adelina said.

"I think that will work. Write it down and carry it with you. You need to keep a list of the stores that you visit. You also need to keep a list of each store's donation. We will need help to pick the donations up. What about the cans for change?"

"We could leave it there if they agree to place it on their counter," Carlos said.

"You will also need a list of stores that accepted the cans. You need to go around every week and collect the money from the cans. At least two of you should go together to do that. You can bring the money to me. I will open a special account at the bank. Will one of you go with me when I do that? We need a treasurer. Who wants to volunteer?" Mrs. Carrillo asked.

"I will," Marcos said.

"It will be your responsibility to count the money with me and to report how much we have at each meeting. You will report both the grand total and the amount earned that week. Does that make sense?" Mrs. Carrillo asked.

"Yes. I can do that."

The plans were taking form. The presentation was ready; it was time to start visiting the stores. Carlos had to work to get the cans ready to distribute. The Flower Festival was growing closer. The three of them made plans to begin visiting the stores the following week.

As the days passed, Marcos wondered if their plans would really work. Could they really build a biological corridor and help the wildlife survive? The reporter seemed to like their project. Her article would be in the paper next week. Mrs. Carrillo thought it would be good to wait until the article came out before visiting the stores. What will happen if people think it is a silly idea? What if nobody contributes? What if nobody helps? What will we do? He felt a little sick to his stomach.

Finally, the newspaper article came out. In fact, it made the front page. The headline read, 'Kids Build Biological Corridor.' Marcos thought, 'Not yet. We haven't done it yet.' But, it turned out okay. People seemed to really like the idea of building a highway for the wildlife. They even liked the idea that kids were working to build it. Now, if they would just donate to the project, all would be well.

The time had come, the three of them set out to pitch their story to the local stores. The first store was a bakery. They stopped in front of the store, looked hesitantly at each other and slowly walked through the door.

Adelina began, "Hi. We are here because lots of animals and birds have gone extinct."

Interrupting her, the store manager said, "I know who you are. You are one of those kids trying to build that highway of trees for the birds. Isn't that right?"

"Yes. We wondered if you could help us by donating some baked goods to sell at our booth at the Flower Festival."

"Sure I can. What would you like? I could give you a couple of cakes and three dozen cookies. Would that help?"

"Yes. That would be great. Could you also put this can on the counter? People can put their spare change in there to help us buy trees," Carlos said.

"I could do that. I like the way you decorated the can. Is there anything else?"

"We need you to sign here. It says that you will donate two cakes and three dozen cookies. We will pick them up the day before the festival. I will come back every week to collect the change from the can. Will that be okay for you?" Carlos asked.

"I think that will work. Here's a cookie for each of you. I think what you are doing is great. I am so glad to see someone helping our native species to survive. Thank you for your good work. Is there anything else?"

"No, sir. Thank you, sir."

So it went, store by store. Each wanted to help. Each donated something. And they all agreed to take a can. At the end of the week, they had fifteen stores signed up.

Marcos began to wonder if they could sell all the baked goods. He also needed help to collect the money from the cans. Adelina and Carlos agreed to help. They would have to do the collections on Tuesdays. Then, they could report how much they made at the Wednesday meeting and give the money to Mrs. Carrillo.

It was just a day before the Flower Festival, and the baked goods needed to be picked up and delivered to The Nature Club's booth. Adelina's parents volunteered to use their truck to do the pick-ups and delivery.

Marcos had been tracking the money from the cans. In three short weeks, the cans had brought in enough money that Marcos thought they could buy a few trees, although he wasn't sure what a tree cost. Still, he felt sure they could buy at least two or three trees. It was a start.

Pedro and Marcos drove to pick Julio up to work in the booth. It was sugarcane cutting time. As they drove, Marcos watched the men in the fields cutting the cane with their machetes. It was like a dance, so rhythmic – the rising and the falling of the machete. The farm laborers lived near the fields in small brightly painted houses. After the cane was cut, they loaded it onto carts. At one time, the carts were pulled by oxen. But, now they were pulled by small tractors.

There was still one old farmer who used the oxen. His cart was beautifully designed and built by local artisans. Costa Rica is famous for its colorful carts. The carts are still made the same way as in the old days. Many of the carts are orange and blue with flowers of many colors placed all around them. The carts are a part of Costa Rican history.

Marcos knew that when the cane had been cut, then the fields would be burned. The sky would fill with gray smoke from the fires. The ash from the fires provided needed fertilizer for the next crop of cane. The smell of smoke was everywhere. At first, the smoke smelled good, but as the days passed it made him nauseous.

The sugarcane fields seemed to meld into the pineapple fields. Pineapple was funny, because there was always ripe pineapple. It didn't really have a season like other crops. It was how they planted it. They rotated the planting to ensure that some was always ripe. Pineapple is a short plant with thick strong leaves. The fruit extends upward out of the center of the plant with sharp edges. To harvest it, they cut the pineapple fruit out away from the plant. They often cut it when it was still a little green. It would ripen away from the plant. Marcos had fresh pineapple for every meal. His mouth watered just thinking about it.

They finally reached Julio's house. "Hi Grandpa," Marcos called.

Julio was ready to go. He climbed into the truck, and they took off to the festival. "Do you think there will be a lot of people at the festival?" Marcos asked.

"There are usually quite a few. I just hope they want something sweet to eat. We have a lot of baked goods to sell." Grandpa smiled.

They rode quietly through the pastures. Julio raised Brahmans. They were really large white cows originally imported from India. In Costa Rica, Brahmans were raised for their meat. Another type cow, Jerseys, was raised for milking.

"I remember when the government offered us a cow if we would clear a hecter (2 ½ acres) of land. The cow and the land were free to all those who helped clear the forests. Now, we are trying to raise money to replant the trees to save the wildlife. Life is truly strange. There were many people who took advantage of the government's offer. Everyone thought it was a good idea and an opportunity to earn a living for their families.

"But now, looking back, it wasn't such a good idea. Some of our wildlife became extinct as a result of the 'slash and burn' policies related to the forests. And, even more forests are threatened now. Many people still don't understand the consequences of destroying the forests," Grandpa Julio said.

"How do people know what the consequences will be?" Marcos asked.

"It is hard to imagine the consequences of an idea or policy. People think it is good at the time, but they can't see the future. New ideas are important. They can be solutions to current problems. We just have to think through all of the possible outcomes and unintended consequences.

"No one intended to hurt the wildlife. In fact, no one considered it. People needed a way to make a living, so they cleared the forests. The loss of wildlife was an unintended consequence. Now, we understand how important the forests are to preserving the wildlife. So, we are trying to develop the corridors, the highways," Grandpa shared.

"We are here. I will park the truck. Then, we can find the booth. Do you two have everything you need?" Pedro asked.

"I think so."

They found the booth. Adelina and her parents were already there. Mrs. Carrillo was placing the first round of baked goods out for sale. People were beginning to arrive. It was going to be a great day.

It was a grand festival. People were waiting at the gate for it to open. The kids were excited to get to the carnival rides. There was a merry-go-round, a small Ferris wheel, and bumper cars. Parents caught up on the local news with neighbors and enjoyed the festival foods. There were lots of local craft booths – a little something for everyone.

The Nature Club booth stayed busy. People stopped by all day long to buy the goodies. They wanted to hear all about the trees and to lend their support.

Adelina and Marcos worked at the booth all day both days. They were excited by all the sales. Julio worked during both days too.

Chapter 5

At the Wednesday meeting, Marcos presented the earnings from the booth at the Flower Festival. It was a quite a lot of money and the club members were excited. "How many trees can we buy?" someone asked.

"I'm not sure," Mrs. Carrillo replied. "We should invite Molly to talk to us now. We also need a place to plant the trees once we have them. We haven't asked anyone about a location. I will invite Molly to our next meeting."

The meeting broke up. Adelina, Carlos, and Marcos walked toward home together.

"Does it seem like the volcano is smoking more than it was?" Carlos asked.

They all looked in the direction of the volcano. It had been smoking for so long now that it just seemed normal. No one thought much about it any longer.

"I think you are right, Carlos," Marcos agreed. "It is smoking a lot more. I wonder what the news report will say. I will turn it on as soon as I get home."

"I sure hope we can buy enough trees to form a highway," Marcus said. "We made a lot of money – well, a lot more than I ever thought we would."

"Me too," Adelina agreed. "I thought we would sell a few cookies and cakes, but I didn't think so many people would give us baked goods. I really didn't think we would sell them all after I saw how much there was. Your grandpa really helped us a lot, Marcos," Adelina added.

"We have also made a lot of money from the cans that you put out Carlos. Everything we have tried has worked. I don't know how many trees it will buy, but you should feel proud of the great ideas," Marcos said.

As they were walking, Julio saw them from the truck. "Do you want a ride?"

"Sure, Grandpa. Can you take my friends home too?"

"Yes, of course, get in. You kids should be really proud of yourselves. You have done a wonderful job of raising money. When are you going to buy the trees? When will you plant them?" Julio asked.

"We don't know, Grandpa. Mrs. Carrillo is inviting Molly to our next meeting to help us with that."

"Could I come to your next meeting?" Julio asked.

The kids looked at each other. "I think so, Grandpa, but I should ask Mrs. Carrillo. Why do you want to come?"

"I have some ideas that I would like to share. Please ask Mrs. Carrillo if it would be okay for me to come," Julio said.

Julio dropped each of the kids off at their homes. He couldn't stop thinking about building the biological corridor and saving the wildlife. He thought that it would help Costa Rica to become the way it was when he was a kid many years ago. The whole country seemed to be covered in forests back then. He hoped Mrs. Carrillo would let him attend the meeting. He had something important to share.

Marcos hurried to listen to the news report about the changes in the volcano. It was still at an 'advisory' level. But the reporter said the changes in the amount of smoke were a concern. The scientists were considering moving it to a 'watch' level.

The next day at school, Marcos asked Mrs. Carrillo if his grandfather could attend the meeting.

"Marcos, why does he want to come to our meeting?"

"I don't know, Mrs. Carrillo, but he said it was important. He has some ideas that he wants to share. He didn't tell me what they were."

"Okay, Marcos, sure, you invite him to our next meeting. He worked very hard for our booth. I'm sure his ideas will be helpful."

Marcos broke into a big smile. "I'll tell him right after school."

"I am sure you have all noticed that the volcano is smoking more now. The news report said that they are considering moving it to the level of 'watch'. There still isn't anything to be worried about. But, it is good to know that the scientists are keeping a close watch on it. I hope you all have a good night," Mrs. Carrillo said.

As soon as school was out, Marcos called Julio on his cell phone. Julio had a cell phone, but he didn't like it. He just kept it for emergencies. When he heard it ring, he picked it up and saw it was Marcos. "Why are you calling, Marcos?"

"The teacher said you could come to the meeting."

"That's great! I'll be there. Is Molly coming?"

"Mrs. Carrillo said that she is coming. She said she was pretty excited about our project."

"Good. I'll be there. Thank you, Marcos. See you Wednesday."

"Bye, Grandpa."

Marcos was excited about all the good things that the Nature Club was doing. He wondered if it would continue after they planted the trees. Maybe everyone would feel that the mission was accomplished. He hoped that wouldn't be the case. He hoped that they could continue to support the environment.

On his way home, he began thinking about the volcano again. It had been moved to a 'watch' level. It was really smoking a lot.

Mrs. Carrillo started the meeting. "Let's get started. I want to share a little more information about volcanoes before we introduce our special guest today. A volcano is like a mountain that opens down to a pool of liquid rock. This is called magma. It is far below the surface of the earth. So, when a volcano erupts, it spews gases and liquid rock.

"So, magma is the name of the liquid rock that is inside the volcano. When the volcano erupts, the liquid rock is called lava. The temperature of lava is 1300 to 2200

degrees Fahrenheit. It glows red hot to white as it flows down the sides of the volcano. Think about how hot that is. Your mothers bake cookies at about 350 degrees. Lava is at least four times as hot as her oven.

"All of this may sound a little scary, but it is important to understand how a volcano works. Ours is still smoking. The scientists will tell us when it reaches a 'warning' level, so we can be safe. I know that it is a lot to think about, but try not to worry.

"Now, we have a special guest today. Molly works to save wildlife here in Costa Rica from extinction. You remember she visited us once before. I have told her about our project. I'll let her tell you the rest. Let's give her a big welcome."

As everyone clapped to welcome her, Molly smiled and walked to the front of the room.

"Hi! Mrs. Carrillo has told me that you have worked very hard to raise money to protect our wildlife. You have done a wonderful job. I understand that you want to use the money to create a biological corridor. I would like to help you select a location. Hopefully, it will be the first of many. We can also determine the types of trees we want to plant to ensure maximum benefit to the most species. We want to include the three-wattled bellbird. He is my favorite endangered species. I want to save him. He needs native avocado trees.

"Let's start with location. Of course, we need the help of some farmers. We want to use land they normally would not use to plant crops. It should be land that would not be used for grazing cattle either. We want the land to be an open stretch between two forested areas that are too far apart for the wildlife to make it from one to the other. Do you have any ideas of where we might find such a piece of land?" Molly asked.

The kids just looked at each other. They would need to think about this. Then Julio asked, "Would a piece of my land work for this project?"

"Maybe. Where is your land?" Molly asked.

Julio explained the location of his farm. He shared that the land was open. Most of the trees had been cut.

"Great, I think it might work. Is there a piece that isn't worth too much to you?" Molly asked.

"There is just such a piece. But, if another piece of land would work better, I would be happy to give it to the project. I have seen so many species disappear. I want to help save those that are left, including your bird, Molly," Julio said.

"Could I visit your farm tomorrow?" Molly asked.

"Yes, of course. I will be there all day. Honk your horn when you arrive. I will be able to hear it from the field. I will be working with the cattle tomorrow."

"Great! I will be there."

"How many trees can we buy? How much is each tree?" Carlos asked.

"It depends on the type of tree. You may have enough for a single stretch of trees that could bridge between forests. We could make it wider when we raised more money. Could I come back next week? I will have more answers by then," Molly said.

"Yes. We would welcome you can come back next week," Mrs. Carrillo said. "Or anytime you can come."

Chapter 6

Molly visited the local plant nurseries. She met with the managers and explained her mission. She told them that the Nature Club was earning money for the trees to create the corridors for wildlife. They seemed excited about the project, but they were not willing to help. It was discouraging, but Molly kept trying to find one that could help.

Finally, Elena, the manager of one said, "That is interesting. I heard that people were working to restore some of the forests. But, you are the first person that I have met who is actually working on it. I am impressed with the idea. I would like to help the kids. The nursery sells native trees. The only way that I can help is to let you have the trees for what they cost me. I have to recover the amount that we actually pay for them. If you pay my cost, it will bring the price of the trees down considerably. Would that work for you?"

"That would be wonderful," Molly replied. "I must admit that I have tried other nurseries, but they were not interested in our project. So, I am really grateful to you. What trees would you suggest for this project? We need to make sure that they are native to Costa Rica."

"I would suggest a diversity of trees," Elena continued. "It might be better for the needs of a variety of wildlife. I would suggest the Guanacoste tree which is Costa Rica's national tree. It is a large tree and can be used for shade. I would also suggest the Ron-ron tree. It is beautiful with reddish leaves and gives off a very pleasant odor which may attract certain birds and insects. The vanilla tree is found everywhere here. People even use it for living fences. That would be an easy to grow tree. The almond tree is important, because it provides a fruit in the form of a nut. The

Machete tree has a beautiful flower and is popular. I would start with a mix of these trees."

"That sounds like a good idea and a balanced way to start the corridor. How about including the avocado tree? We need it for the three-wattled bellbird," Molly said.

"We could include the avocado tree, but it will be a little more expensive. Do you still want it?" Elena asked.

"Yes, it is important to our project. We are hoping to save the three-wattled bellbird. Would you come to the next meeting of the club? You could talk about the trees. I think your plan sounds like the right way to approach this project," Molly said.

"I would love to come. I will put it on my calendar. I'll see you next Wednesday. This will be fun."

Molly felt really good about her meeting with Elena. She knew that it was important to have a variety of trees. Different species of wildlife needed the different trees. Next on Molly's 'to do' list was to visit Julio's farm. She needed to determine if he had a piece of land that could be used as a corridor. She decided to go straight from the nursery. She called Julio to make sure it was a convenient time to visit. He told her to come right over. She did, and honked when she arrived. Julio was working in the field. He came right up.

"Hi, Molly. It is good to see you. Are you ready to see your first corridor?" Julio asked smiling.

"I am ready." Molly loved his enthusiasm and willingness to help. She had expected that finding a piece of land would be difficult. So, Julio's generous offer was a pleasant surprise.

They saddled up the horses. It would be easiest to see the stretch of land Julio was thinking of from horseback. They arrived at the location in about a half hour. It was an area with a wide ditch that was covered in rocks and grass. It was no good for farming.

"This looks good. Do you see that stretch of trees over there? This is close enough; the birds could fly between them without needing to stop. There is not one on the other side, but that could be our next project. If we built that one, they could get to the Cloud Forest. Are you sure you are comfortable donating this land to trees? We can't pay you anything, Julio," Molly said.

"I know. I want to help. I can't use this land for anything else. Truthfully, there is another piece over there where you were pointing. You can have that too," Julio shared.

"That is perfect. That will make it easier for the wildlife, especially the three-wattled bellbird. Will you come to the Nature Club meeting on Wednesday?"

"Yes, I would be happy to come."

"I wish other farmers felt like you do. Many of them don't want to think about the forest or wildlife. I am worried for the future of the Cloud Forest. But, maybe you will be an example for the others. Thank you so much for sharing your farm. I will see you on Wednesday." Molly waved as she left.

Molly called Mrs. Carrillo to let her know that Julio and Elena would be attending the meeting. "It should be a good meeting. The kids will be excited that we are so close to planting trees. I wonder if I should invite Martin from the Biology Club at the high school. They have volunteered to help plant the trees once we purchase them," Mrs. Carrillo said.

"I think that would be good. This meeting will bring the whole project into reality for everyone. I think Elena will bring examples of native trees that will work for this project. We may need extra time for this meeting."

"I will send out an email to the parents to let them know that the kids will be late. I'll get everything ready for the meeting. Thanks for all your help, Molly. I don't think we could do it without you."

Wednesday came quickly. Mrs. Carrillo called the meeting to order. Julio, Martin, and Elena were in attendance. She introduced each of them to the class.

"I would like to begin today with a presentation from Elena who works at a plant nursery. She will be selling the trees to us. She is going to share some information about the trees. She is giving us quite a discount, so we need to welcome her and also thank her. Let's give her a big hand."

"It's nice to be here. I wanted to talk with you today about a couple of native trees. I will start with the idea that it is best to plant a combination of trees. They can be divided into hardwoods and fruit trees. Different species of insects, birds and animals prefer different trees and plants. I brought a couple of small trees with me to show you. First, I have the Almendro (almond) tree which is native to the rain forests. It forms a yellowish bark which is distinct from the other trees. I also have a Machete tree which has orange blossoms in the spring. It is called the Machete tree by the locals because the flowers resemble a machete. My last tree is the Guanacaste tree which is our national tree," Elena said.

"They are sure small," Marcos said.

"They are small, but they will grow quickly once they are planted. You need a lot of trees to begin this project. The smaller trees are less expensive, so it is better to have more trees," Elena said.

"Trees and plants do grow fast. We planted one in our yard. It was big by the next year," Adelina said.

"You will need about 200 trees for the stretch of land that Julio has donated. It is a good sized piece of land. He has another piece that he is willing to share with us in the future. The two would complete a corridor between two forests. It would be wonderful if we could continue with the project after the first corridor is completed," Molly said.

"Thank you, Elena. Now, I also want to thank Julio for giving us the land. We have Martin with us today too. He will tell us how to proceed with planting the trees," Mrs. Carillo said.

"The high school Biology Club has recruited quite a few volunteers to help plant the trees. We can begin in two weeks. We will have a crew of ten students ready with shovels and tools. We will meet you at Julio's farm at eight o'clock on Saturday morning. We need you to bring the trees. We will also need a couple of you to bring items to workers. We call these people "runners". They run for whatever is needed. It would also be good if we could have some food and water. Has anyone thought of that?" Martin asked.

"Martin, we have not considered food or water for the crew of planters. But, I think we can get help to provide that for everyone who is working," Mrs. Carrillo said.

"We also need planting soil from the nursery to give the trees the very best chance to survive. Has that been figured into the cost?"

"I can volunteer help from the nursery with the necessary additives to the soil. I think we can do that for a very low cost," Elena said.

"We can get "runners" from the Nature Club. I will volunteer to help," Marcos said.

"I think that will do it. Just think, the next time we meet, we will be planting trees to build the corridor on Julio's farm. The dream is becoming a reality," Martin shared.

"Thank you Martin. I think that will do it. We have work to do. I'll see you all in two weeks," Mrs. Carrillo said.

During the time they were preparing to plant the trees, the volcano developed a fissure vent allowing lava to escape from the volcano without an eruption. It created a river of red flowing down the side of the volcano. The town held a small celebration. There would be no major eruption. The fissure vent allowed the release of the built up pressure inside the volcano. Everyone in the community breathed a big sigh of relief.

Chapter 7

The day for planting finally arrived. Everyone met at Julio's farm just as they planned. Elena donated a few extra trees to ensure that the corridor was large enough to assist the wildlife. The Biology Club was out in force. Twenty students showed up with shovels instead of the expected ten. Parents and families came with enough food to feed the kids for a week. Julio couldn't stop smiling. He was so happy to have the biological corridor on his land. He knew there would be more corridors to follow.

Parents brought shovels on Sunday to help the Biology Club finish. The trees were all planted over the course of the weekend. Julio invited all the students to come whenever they wanted to check on the plants. He checked them every day to make sure they were okay. Some of the students from each of the clubs visited the farm often to watch the trees grow.

It seemed to take forever, but finally the trees started putting out new growth in the form of leaves and branches. The kids were thrilled. Nearly all of them were growing. It was going to work. The wildlife would have a highway through Julio's land.

Julio was nursing a couple of the trees. They just didn't want to grow. He asked Elena about them. She gave him some special plant food. He gave it to the trees as she had directed. But, they were small. It didn't seem to help. What could he do? The trees were expensive. He didn't want to lose any.

Carlos and Marcos dropped by one day. "What is wrong with these trees?" Marcos asked.

"I don't know. I've asked Elena. She gave me some special tree food, but it isn't helping," Julio said.

"What can we do?" Carlos asked.

"I don't think there is much to be done. I think we just have to wait and see what happens," Julio said.

A car drove into the farm. They looked up. It was Molly. She had come to visit the trees too. She walked over to them.

"What's going on?" Molly asked.

"It's these trees. They are not growing," Marcos said.

"I have tried everything. I've spoken with Elena, but nothing is working," Julio said.

"Once, a long time ago, my grandmother was having a problem with some of her plants that wouldn't grow. Just like the trees, nothing helped. One day, she started digging her leftover vegetables into the ground around the trees. She used things like potato peels, the parts of vegetables that we discard," Molly said.

"Did it help?" Marcos asked.

"Yes, it did. It took a couple of weeks, but those plants became the biggest and strongest in her garden."

"It's worth a try. Where can we get enough of the vegetable discards to nurture the trees?" Julio asked.

"I can help. Mom will give me hers," Marcos said.

"Me too," Carlos said. "We can ask the other kids too."

"We will ask the Nature Club to help."

"Good. We will see if it works. It can't hurt them. I don't know of any other remedy," Molly said.

"I'm willing to try anything," Julio said.

Marcos and Carlos presented the problem to the Nature Club. The kids volunteered to help. There were so many volunteers that they decided to take turns giving their veggie discards to Julio.

The discards were worked into the soil every day. It was about two weeks later when Julio noticed new leaves on the tiny trees. He was thrilled. He called Marcos to let him know that the trees had started to grow.

Marcos shared the news at the next club meeting. Everyone was excited.

By the time fall came, the trees were taller and stronger. The birds could sit on their branches. It was hard to believe that nearly a year had passed since the Nature Club formed. The biological corridor was built, and birds were sitting in the trees. Julio thought he spotted a three-wattled bellbird. But, it was so far away that he couldn't be sure.

The Nature Club continued to meet. Molly was visiting. She had a proposition for the club.

"The first biological corridor is built, and some of the wildlife has found it. As the trees grow, more birds will find and use them. A few epiphytes have started to grow on the trees. They are the plants that live on other plants. I have noticed a few orchids and some mosses. It is a good indication that the trees are healthy. You should all feel very proud of your accomplishments. You led the way for the whole community on this project. I have a special award for each of you.

"Mrs. Carrillo, would you invite the parents into the classroom now?"

While Mrs. Carrillo was getting the parents, Molly kept talking. "I am hoping that together we will build more corridors. The Nature Club is very important to this project."

Mrs. Carrillo walked to the library where the parents were waiting and invited them to the classroom. As they gathered in the back of the room, Molly began her presentation.

"Each of you has worked hard on this project. Each of you contributed to making a difference for Costa Rica. You took the first step to change a mistake from the past to create a better chance for the future of wildlife. Each of you deserves credit for your part in this project. I have a special award to thank you for your participation." Everyone clapped as each of students went up to accept their award.

"I want to give special credit to Marcos. He brought the problem to the school. He initiated the process that built the club, secured the trees, and culminated in establishing a new habitat, a new highway for migrating birds. Marcos, I have this very special award for you. Congratulations! Thank you for being an excellent leader in this project. Please accept this award."

Molly handed him a large wooden plaque engraved with his name and a short message. It read, 'For Marcos, who had the courage and foresight to lead the Nature Club to build the biological corridor.' Marcos smiled as he thanked Molly for the award.

After the meeting, Pedro put his arm around Marcos' shoulder. "Marcos, you have made me very proud. I don't like to tell you that, because I am afraid you will stop trying. But, I have to tell you now. I am so proud of you today that my eyes are filled with tears." He gave Marcos a big hug.

Marcos smiled through his tears. "I won't stop trying, Papa. I like to do things that help people and animals."

With that, they went home where a celebration with cake and ice cream was waiting for them.